SALVADOS DE LA EXTINCIÓN

LOS ANFIBIOS

BLACK RABBIT BOOKS

JOANNE MATTERN

TABLA DE CONTENIDOS

1

Rana leopardo de Chiricahua

¿Oyes ronquidos? ¡No! Es una rana leopardo de Chiricahua que hace ruido.

Estas ranas viven en Arizona y Nuevo México. Necesitan agua para poner sus huevos. Sin embargo, ha desaparecido mucha agua. Los ganaderos drenaron los estanques para poder criar ganado. A finales de la década de 1990, solo quedaban 80 lugares donde las ranas podían vivir.

La gente trabajó en conjunto para salvar a las ranas. Construyeron nuevos tanques de agua. Los zoológicos criaron las ranas. Luego las liberaron. Hoy hay más de 10,000 de estas ranas en estado silvestre.

¿Sabías que...?

Estas ranas también enfrentan problemas con las ranas toro que se comen sus huevos.

2

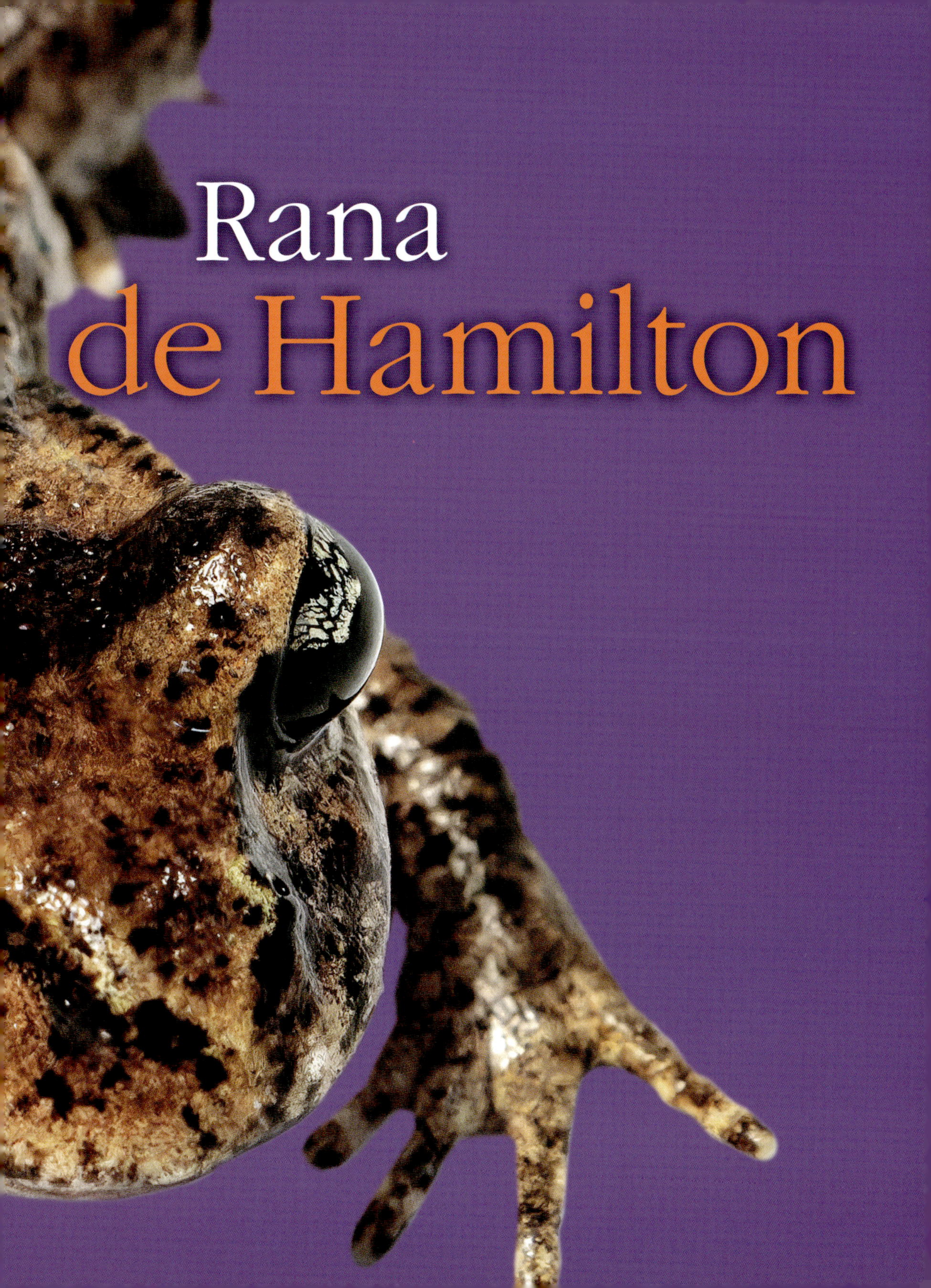
Rana
de Hamilton

La rana de Hamilton es diminuta. Mide menos de 2 pulgadas (5 centímetros) de largo. Estas ranas solían vivir por toda Nueva Zelanda. Sin embargo, muchos depredadores las mataron. En 2004, solo quedaban unas 300.

Los científicos trabajaron arduamente para salvar a estos diminutos anfibios. Crearon un nuevo lugar para que vivieran las ranas. Luego llevaron 80 ranas a su nuevo hogar. Hoy, muchas ranas viven en dos áreas de Nueva Zelanda.

Piensa en esto

¿Por qué crees que los depredadores representan una mayor amenaza para una especie que para otras?

Salamandra
Hellbender
3
10

Las salamandras Hellbender viven en arroyos correntosos del este de Estados Unidos. Necesitan agua para sobrevivir. Pero se construyeron represas en sus hogares. La gente contaminó el agua. En 2011, los animales casi habían desaparecido.

Luego la gente intervino para ayudar. Limpiaron los ríos y arroyos. Hicieron de los ríos lugares saludables para que vivieran las salamandras.

Los zoológicos también ayudaron. Criaron algunas salamandras en cautiverio. Luego las devolvieron a su hábitat natural.

Piensa en esto

Los anfibios pueden ser una pista de si el agua está limpia. ¿Por qué crees que es así?

Rana de montaña de patas amarillas

4

La mayoría de las ranas viven donde hace calor. ¡Pero no la rana de montaña de patas amarillas! Vive en las alturas de las frías montañas de California.

Antes, estas ranas eran comunes. Luego la gente quiso pescar truchas. Se añadieron muchos peces a los lagos de las ranas. Las truchas se comieron los huevos de las ranas. Pronto apenas quedaban ranas.

La gente intervino para salvar a las ranas. El gobierno eliminó las truchas. Los zoológicos criaron a las ranas. Luego las devolvieron a los lagos. Hoy, estas ranas se encuentran bien.

¿Sabías que...?
Estas ranas **hibernan** en el fondo de los lagos durante el invierno.

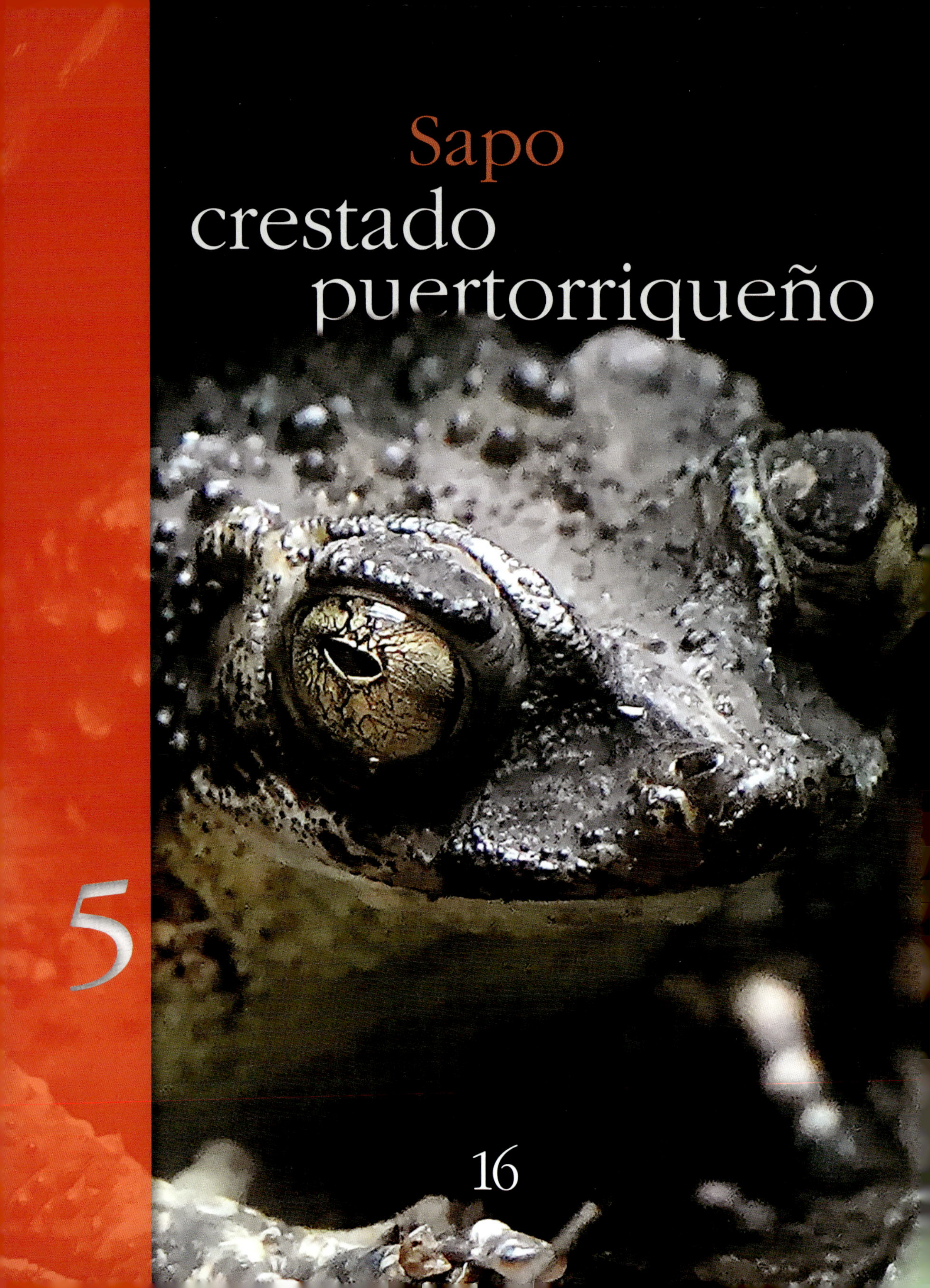

Sapo crestado puertorriqueño

5

Los científicos vieron por primera vez al sapo crestado puertorriqueño en la década de 1860. Luego, el animal desapareció. Durante años, la gente pensó que se había extinguido. Luego, se avistaron algunos en estado silvestre. Los científicos tuvieron la oportunidad de salvar al sapo.

Capturaron algunos de los sapos. Los enviaron a zoológicos de Estados Unidos y Canadá. Allí los criaron. Luego, los zoológicos liberaron miles de renacuajos en la naturaleza. Este sapo todavía está en peligro. Pero ahora tiene la oportunidad de sobrevivir.

Piensa en esto

¿Por qué son importantes los zoológicos para salvar especies en peligro de extinción?

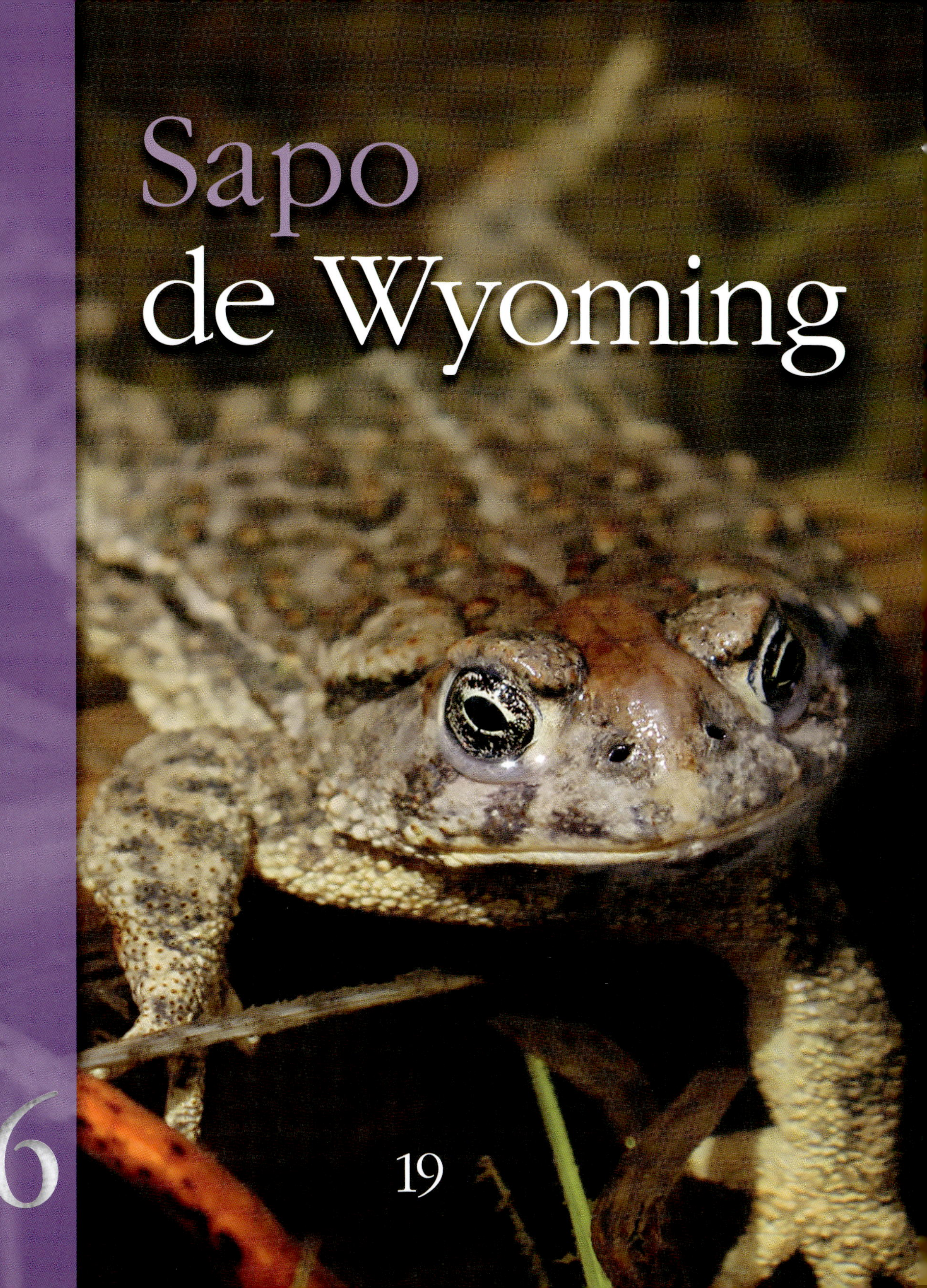

Sapo de Wyoming

6

El sapo de Wyoming vive en un solo lugar del mundo. Pero las enfermedades y la pérdida de **hábitat** lo pusieron en peligro. Los científicos estaban muy preocupados. En 1994 capturaron los últimos 10 sapos.

Estos fueron criados en zoológicos. Luego, en 2023, consiguieron un nuevo hogar. El gobierno creó un **refugio** para ellos. Está cerca de Laramie, Wyoming. Ahora estos sapos tienen un lugar seguro donde vivir.

¿Sabías que...?

El refugio es el único lugar donde los sapos de Wyoming viven en estado silvestre.

MÁS PARA EXPLORAR

¿EN QUÉ PARTE DEL MUNDO?

OCÉANO PACÍFICO

AMÉRICA DEL NORTE

OCÉANO ATLÁNTICO

EUROPA

ASIA

OCÉANO PACÍFICO

Puerto Rico

ÁFRICA

OCÉANO ÍNDICO

AMÉRICA DEL SUR

AUSTRALIA

Nueva Zelanda

RANA DE MONTAÑA DE PATAS AMARILLAS
California

RANA LEOPARDO DE CHIRICAHUA
Arizona y Nuevo México

SAPO DE WYOMING
Sur de Wyoming

SAPO CRESTADO PUERTORRIQUEÑO
Puerto Rico

SALAMANDRA HELLBENDER
Este de Estados Unidos

RANA DE HAMILTON
Nueva Zelanda

MÁS PARA EXPLORAR

DATOS FANTÁSTICOS

Los incendios forestales a menudo destruyen la hierba alta donde anidan las ranas leopardo de Chiricahua.

Las ranas de Hamilton no tienen orejas ni cuerdas vocales.

Las salamandras Hellbender están cubiertas de baba. Parecen venenosas, pero no lo son.

Las ranas de montaña de patas amarillas se tumban sobre rocas soleadas para calentarse.

El sapo crestado puertorriqueño es el único sapo nativo de Puerto Rico.

Un sapo de Wyoming puede liberar veneno desde la parte superior de su cabeza.

MÁS PARA EXPLORAR

COMPARACIONES INTERESANTES

Estos animales están reapareciendo. ¿Cuántos viven en estado silvestre?

Rana leopardo de Chiricahua
alrededor de 10,000

Rana de Hamilton
alrededor de 300

Salamandra Hellbender
alrededor de 915

Rana de montaña de patas amarillas
alrededor de 200

Sapo crestado puertorriqueño
entre 1,000 y 3,000

Sapo de Wyoming
alrededor de 1,500

MÁS PARA EXPLORAR

RECURSOS

Glosario

anfibio Un animal de sangre fría que puede vivir en la tierra y en el agua.

cautiverio El estado de estar retenido en un lugar, como un zoológico.

depredador Un animal que come a otros animales.

en peligro de extinción A punto de desaparecer.

hábitat El lugar donde crece o vive una planta o un animal.

hibernar Pasar el invierno durmiendo o descansando.

refugio Un lugar que proporciona cobijo o protección.

Índice alfabético

TOP RANK es publicado por Black Rabbit Books, P.O. Box 227, Mankato, MN, 56002. •

• Top Rank un sello de Black Rabbit Books • Diseñadora de la serie: Danny Nanos • Diseñadora del libro: Jason Knudson • Fotografías © Alamy Stock Photo/John Cancalosi, 5, 21, 23, Kevin Schafer, cover, 1, Natural History Collection, 15, Tom Grundy, 14; © Andy MacDonald/CC BY-NC 4.0, 9, 21; Flickr/Bill Dorsey, 16, Charles (Chuck) Peterson, 4, Shannon Buttimer, 20; Minden Pictures/Joel Sartore/Photo Ark, 2, Pete Oxford, 10–11; Oscar Thomas, 6–7; Shutterstock/Nastya Tsvyk, cover, 1; USGS/Public Domain, 13, 21, 23; Wikimedia Commons/brian.gratwicke, 12, 21, 23, Jan P. Zegarra, 18, U.S. Fish and Wildlife Service Southeast Region, 17, 21, 23, USDA Forest Service, 19, 21, 23 • Impreso en India

Library of Congress Cataloging-in-Publication Data: Names: Mattern, Joanne, 1963- author. | Title: Los anfibios: salvados de la extinción / by Joanne Mattern. | Other titles: Amphibians. Spanish | Description: Mankato, MN: Top Rank is an imprint of Black Rabbit Books, [2026] | Series: Salvados de la extinción | Includes index. | Ages 8–11 | Grades 2–3 | Identifiers: LCCN 2024054666 (print) | LCCN 2024054667 (ebook) | ISBN 9781644669686 (library binding) | ISBN 9781644669822 (paperback) | ISBN 9781644669969 (ebook) | Subjects: LCSH: Rare amphibians—Juvenile literature. | Amphibians—Conservation—Juvenile literature. | Endangered species—Juvenile literature. | Wildlife recovery—Juvenile literature. | Wildlife conservation—Juvenile literature. | Classification: LCC QL644.7 .M37318 2026 (print) | LCC QL644.7 (ebook) | DDC 597.8—dc23/eng/2025012

Who Plays Online *Games*

5

A lot of people think teenage boys play the most video games. But this is a **stereotype**. The biggest group of gamers is adults 18 to 34 years old. And girls like to play video games too. A 2024 study found 46 percent of gamers identify as female, while 53 percent identify as male. One percent of gamers identify as nonbinary. There are nearly as many female gamers as there are male gamers.

Another study found adult women play more video games than teenage boys. Mobile devices have increased the number of female players. Women prefer playing online games on their smartphones. Some people think women prefer *Candy Crush Saga* and *Words with Friends*. That is another stereotype. Different female players enjoy different games. Some women enjoy brain teaser and card game apps. Others

like playing adventure and other games on their PCs.

About 38 percent of gamers are 18 to 34 years old. They make up the most gamers in any age group. Males tend to like games full of competition and 3D navigation. They often like games with multiplayer functions so they can play with friends. But be careful not to stereotype men—or anyone else. Anybody can like any type of game.

Some gamers use a virtual reality headset to play games.

61 Percent of Americans ages 5 to 90 who play video games.

The average gamer is about 36 years old. • China is the country with the most gamers. • ESRB ratings help people find games that are appropriate for their age.

A teen girl searches for targets in a shooter game.

ESRB The Entertainment Software Rating Board is the ESRB. It rates video games. It began in 1994. There are four ratings. They include "everyone," "everyone 10+," "teen," and "mature." You can use the ratings to find games that are right for you. Games with higher ratings have content that is not appropriate for kids.

What Are the Dangers of *Online Gaming?*

6

Online gaming can be fun. However, players must remember to look out for real-life dangers. When you play online, information is stored in your computer or console. Over time, you add more personal information to online profiles and gaming websites. This information is collected and used to create your online identity.

If a gaming website or network asks you to create a profile, do not add personal information. That includes your last name, address, telephone number, or school. Keep this information private. This will help you stay safe. Your online identity can reveal personal information. Your future employers, colleges, and coworkers could search for your online identity. They could learn more about you. They might find out information you'd rather they didn't.

Never use your parent's personal information online, either. That includes a credit card to make online purchases. If you buy something from an untrusted website, criminals could find the credit card number. Think twice before downloading a new game or cheat code. Some may have a computer **virus**.

Hackers can steal your personal information if you aren't careful.

25 Percent of parents who worry about how much time their children spend online playing games.

Information online could help strangers find you in real life. • About 75 percent of gamers have experienced bullying when gaming. • Regular updates can help keep your device protected.

How Do I Stay Safe *While Gaming?*

7

Online gaming is fun. But it's good to know how to keep yourself safe while you play. First, make sure your gaming device is protected. Install protection, such as anti-virus and anti-**spyware** software. Computer viruses harm your device. Some let other people see your personal information. Ask a trusted adult to help you protect your device.

Most online games require you to create an account. Manage your account with safety in mind. Choose a username that does not give away your identity. You will need a strong password as well. Many accounts include a user image. Instead of a photo of yourself, use an **avatar**. An avatar lets you create your own "virtual self."

Many games allow you to interact with other players. Some games let you text or voice chat. If you use voice chat, **disguise** your voice. Otherwise, it is safer to use

Anti-virus apps protect your computer from viruses.

text chat with strangers. If someone sends you a link, do not open it. It may have a virus. Avoid using a webcam when playing online. And never meet an online stranger in person. Keeping your real-life identity private helps you stay safe.

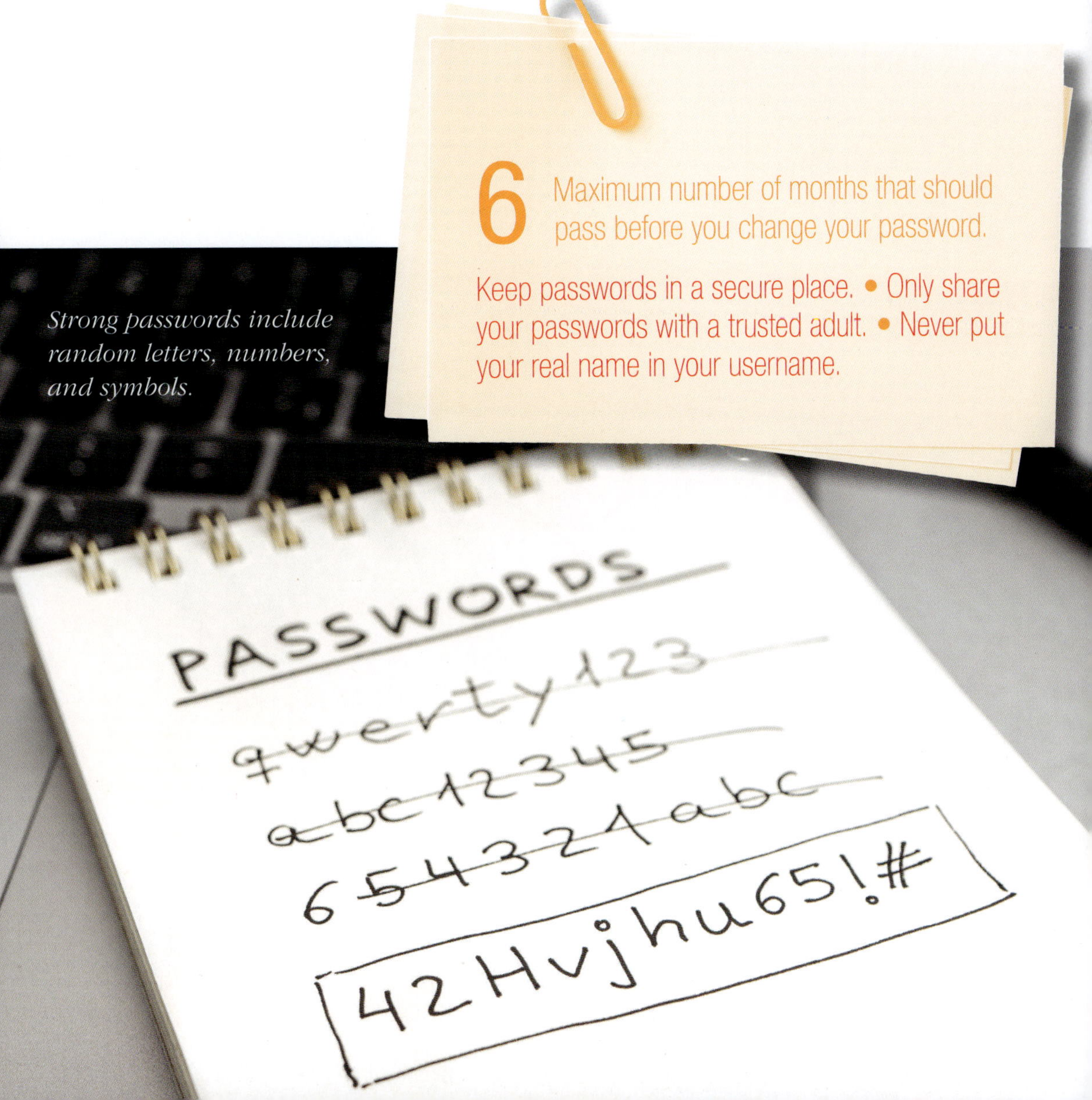

Strong passwords include random letters, numbers, and symbols.

6 Maximum number of months that should pass before you change your password.

Keep passwords in a secure place. • Only share your passwords with a trusted adult. • Never put your real name in your username.

Are Violent Games *Bad for You?*

8

Not all video games are violent. But some have extreme violence. Scientists have studied video game violence. Some people think these games can cause players to become violent in real life. So far, scientists are unsure if this is the case.

One study watched gamers after they played a violent game. It found that it takes gamers four minutes to get rid of violent thoughts after playing. It takes up to ten minutes for their heart rates to return to normal. Some research suggests that violent video games make players less understanding of others' feelings. Other scientists disagree. Some research suggests violent video games give players a way to release their own anger. A 2024 study found 68 percent of gamers play video games to relax. And 67 percent said games helped them have fun. Those seem to be gamers' **priorities**.

More research suggests violent video games can teach players there are consequences for their actions. Interactions in video games may teach players how to interact in the real world. Sometimes, online characters act violently and get away with it. Most of the time, though, there are in-game consequences. Of course, the player is not punished in real life. But the game does help show the kind of consequences there could be. Plus, most gamers understand that video games aren't real.

13 Percent of the games released in 2023 that received an M rating.

Games are rated M for mature. You need to be at least 17 years old to buy them. • In 2023, 23 percent of games were rated T for "teens." • Most games released in 2023 were rated E for "everyone."

GAME LESSONS *The Legend of Zelda* has some fantasy violence. But it still teaches valuable lessons. For example, one part of the game is Zelda's Money Making Game. It teaches about gambling. In the game, you will either lose 10 rupees or earn 50 rupees. But sometimes you lose 40 rupees. This can teach the consequences of gambling in the real world.

Many Call of Duty *games are rated M for violence and strong language.*

Is Online Gaming *Addictive?*

9

Online gaming is different from playing video games offline. Many online games do not have a beginning, middle, and end. This is especially true of MMO games. Because many online games never actually end, some people have trouble turning the game off. It is addicting.

It is easy to lose track of time while gaming. It can get in the way of doing homework. A gamer might lose sleep or not take care of themselves. They may ignore their friends and family. Some people use the online gaming community to **socialize**. They meet people online instead of in person. This can affect someone's social skills. In-game purchases can also be a problem. Gamers spend more and more money to play. All of these are signs of addiction.

Gaming console gift cards are sold in many stores.

Addiction can often begin with a different problem. For example, someone

may be getting bullied at school. Or there might be trouble at home. In these situations, they use gaming to escape. They don't have to think about their problems as they are playing. If you are worried about an addiction, seek the help of a trusted adult. Share any problems you might be trying to avoid.

THINK ABOUT IT Research how game developers create games. How does this affect someone who can't stop playing?

Parents can watch for signs of addiction.

A gaming addiction can turn into a financial burden that is hard to recover from.

7 Average number of hours an American plays video games each week.

Too much gaming can cause eye strain. • Up to 10 percent of people in the United States are addicted to gaming. • Males are more likely to become addicted to gaming. It can affect anyone, though.

How Do Game Makers Keep Players *Coming Back?*

10

A video game researcher's job is to find out what keeps all gamers coming back. The term *gameplay* refers to how a player interacts with a game. Many games have a plot, characters, and a setting. They also feature a **climax** and specific challenges. Gameplay connects your character to those feature

Researchers study the replay value of a game. This shows if players will return to a game more than once. Some games follow a story. In these games, characters start at the beginning. Then they move through the middle and reach the end. This is called linear gameplay. In RPGs, there are hundreds of **side quests**. They often have thousands of non-player characters (NPCs). Many RPGs have nonlinear gameplay. These games do not follow a story line. Players have more choices. They decide how they spend their time in the game.

Video game fans sometimes dress up as their favorite characters.

$214 billion Estimated worth of the video game industry in 2024.

New maps and characters are released often to keep players' interest. • Video game competitions can increase a game's popularity. • The average game developer makes about $115,000 a year.

According to experts, gamers like to complete side quests more than they like to play the actual game. Many players prefer nonlinear gameplay. These games give players more freedom. Players do not need to follow a particular storyline. They get to decide what their character does.

Spider-Man

Famous characters can make a game even more popular.

Mario

What Are the Benefits of *Online Games?*

11

Video games can have many benefits. Online games bring people together. They aren't face-to-face, but they are still socializing. The games can help people feel connected. Players can talk to someone from across the street or the country. Some gamers talk to people around the world. They learn to type and speak other languages. They learn about other cultures.

It is common to have online friends. Playing with others is usually more fun than playing alone. A friend who games with you may be able to help you in the real world too. Parents and kids often play together. They enjoy sharing the activity as a family.

Cooperative games connect people. Players work together for a common goal. They learn how to work as a team and use good communication. Action and adventure games let players discover new places.

Online puzzle games help players feel more confident. They exercise players' brains too.

Some research shows gaming may be good for your brain. It may help your ability to reason and remember things. Games may even help you read maps more easily. They could help you learn to find your way in an unfamiliar place.

Sports video games let gamers play as professional players and teams.

80 Percent of gamers who play with friends.

More than 70 percent of parents who play video games play with their kids. • Many gamers enjoy the social part of gaming most. • Many players get a sense of accomplishment from beating a video game.

Playing a game with your friends can make it more fun.

What Is the Future of *Online Gaming?*

12

Technology changes quickly. This includes online video games. Five years ago, gamers were not playing the same types of games they play today. Five years from now, gaming will probably look different.

The gaming community itself has changed. With the rise of mobile devices, everyone has an opportunity to play. In fact, gamers do not even need a console or a PC. They can download game apps on their smartphones. An app is much cheaper than a console or PC game. And an app is easy to use. You can install it on multiple devices. Gaming on mobile devices is expected to become even more popular.

Many people use apps to play online games.

Video games in classrooms are also growing. These are educational video games. Research shows students become more engaged

when video games are included in lessons. They improve students' problem-solving abilities. They can strengthen critical-thinking skills. They also help kids learn science, math, and history.

Game developers continue to search for new and exciting game features. One is virtual reality (VR). Users wear a headset with a computer. The computer creates a realistic, virtual world. Users can see and hear this new world. VR has become more popular in the past few years. People expect it to grow. There are many predictions for online gaming. But no one can know for sure what the next great game will be.

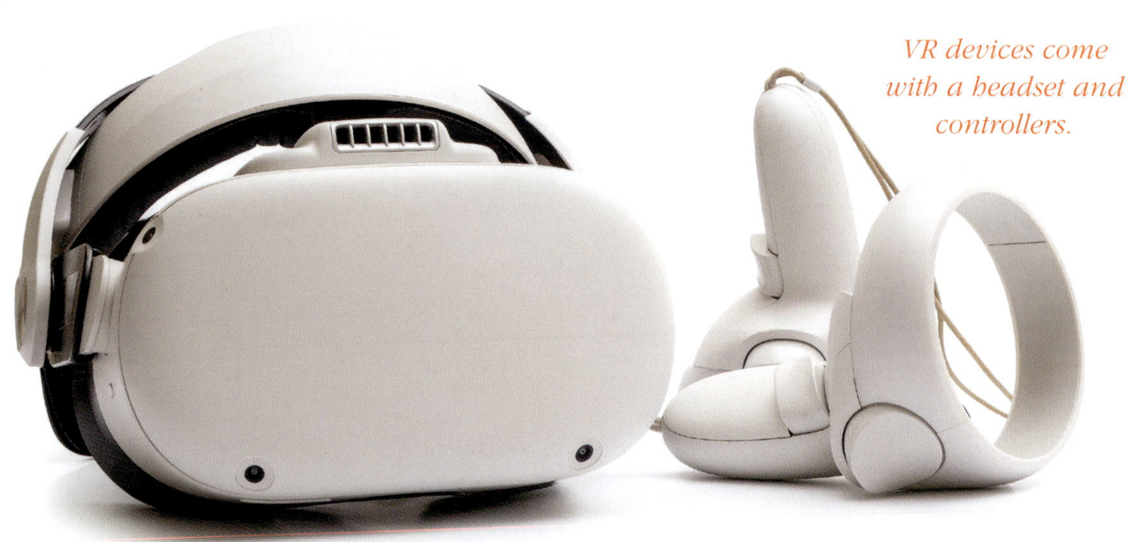

VR devices come with a headset and controllers.

64 million Number of people who used VR in 2022.

About 25 percent of US teens own a VR set. • Some people use VR headsets to watch movies and TV. • VR is expected to become more popular as it becomes cheaper.

Think About It What do you think the future of gaming holds? Do you think VR will ever be more popular than mobile gaming? Why or why not?

VR can let you to experience the wonder of outer space from the comfort of home.

Staying Safe While

Take Breaks

It is easy to play for hours. But sitting for too long is bad for your health. Instead, set a time limit and take short breaks. Stretch your body. Do a quick physical activity. This will help keep your muscles from getting sore. Make sure to stop if you get hungry or tired.

Keep Your Device Updated

Developers are always improving their games. Some updates include security patches. These fix holes in the system that criminals use to find people's information. Download all software updates right away. This will keep your game secure. You can turn on automatic updates. This is in your settings.

Gaming Online

Protect Your Privacy

Use a safe username. Make sure it doesn't include any personal information. Do not include your name, birthday, school name, or city. Change your password regularly too. And check the game's privacy settings. Each game and console are different. Make sure everything is set to private.

Share with Care

Think about who you are playing with. Most online games are open to everyone. People might not be who they say they are. Keep your real-life identity hidden online. Do not share any personal information. If a player makes you feel uncomfortable, block them. Then tell a trusted adult.

Glossary

avatar
A small picture that represents an online user.

climax
An important moment in a game's story where the player's actions and choices lead to the final scene.

cooperative
Involving two or more people or groups working together to do something.

disguise
To change the usual appearance of someone or something so that people won't recognize it.

game console
An electronic device used with a display and controls to play video games.

graphic
A pictorial image created by a computer for display on the screen of a digital device.

mobile
Able to be moved.

realistic
Something that seems very lifelike or real.

side quest
A less important activity in a video game that is done in addition to the main activity or task.

simultaneously
At the same time.

socialize
To talk to and do things with others in a friendly way.

spyware
Computer software that secretly records information about the way you use your computer.

stereotype
A widely believed but unfair image or idea of a particular type of a person or thing.

virtual
Happening online.

virus
A program that is designed to harm a computer.

For More Information

Books

Clark, Katie. *Using Mobile Devices.* Minneapolis: Lerner Publications, 2026.

Gish, Ashley. *MOBA Games.* Mendota Heights, MN: Apex, 2024.

Light, Charlie. *Online Gaming.* Buffalo, NY: Enslow Publishing, 2024.

Websites

Are Video Games Bad for Me?
kidshealth.org/en/kids/video-gaming.html

Is Video Game Addiction Real?
www.pbs.org/video/is-video-game-addiction-real-pz1nmj/

Online Gaming
www.childline.org.uk/info-advice/bullying-abuse-safety/online-mobile-safety/online-gaming/

About the Author

Marysa Storm is an author and editor who has worked on subjects ranging from pets and wildlife, to yoga and ceremonial magic. She lives in the Twin Cities area with her better half and their tortoiseshell cat. She spends her free time reading tarot, working on sewing projects, and rewatching *The X-Files.*

Index

TOP RANK is published by Black Rabbit Books, P.O. Box 227, Mankato, MN, 56002.

• Designed by Danny Nanos • Photographs © Alamy Stock/Photo Westend61, 17, 20; Dreamstime/Adi Pratama, 10, Georgejmclittle, 40, Rafael Henrique, 23; Getty Images/Federico Parra, 35, Stanislav Kogiku/SOPA Images, 15; Shutterstock/Aleksandar Malivuk, 37, ANON KULSUWAN, 2–3, Arsenii Palivoda, 11, Belinda Pretorius, 48, benixs, 18, Cheer Group, cover, 1, Dean Drobot, 5, 39, dennizn, 30, Dmytro Zinkevych, 31, doublelee, 32–33, etonastenka, 6, felixmontana, 9, FUN FUN PHOTO, 2, 36, glenda, 44, Gorodenkoff, 22, HenadziPechan, 12, IhorL, 32, Jolygon, 46–47, Kaspars Grinvalds, 25, KateV28, 28–29, Kitinut Jinapuck, 21, Krivosheev Vitaly, 38, MarinaP, 27, Miguel Lagoa, 13, MMPhoto21, 45, ozrimoz, 24, Phakorn Kasikij, 41, Prostock-studio, 18–19, r.classen, 45, robtek, 16, Shakirov Albert, 7, Summit Art Creations, 44, Tinxi, 4, Trismegist san, 34, 36, vfhnb12, 8, Viacheslav Lopatin, 40, 42–43, Wirestock Creators, 26 • Printed in India

Library of Congress Cataloging-in-Publication Data: Names: Storm, Marysa, author. | Title: 12 questions about online gaming / by Marysa Storm. | Other titles: Twelve questions about online gaming | Description: Mankato, MN Top Rank, an imprint of Black Rabbit Books, [2026] | Series: Tech smarts | Includes bibliographical references and index. | Ages 9–13 | Grades 4–6 | Identifiers: LCCN 2024054682 (print) | LCCN 2024054683 (ebook) | ISBN 9781644668177 (library binding) | ISBN 9781644668498 (paperback) | ISBN 9781644668818 (ebook) | Subjects: LCSH: Video games—Juvenile literature. | Video games and children—Juvenile literature. | Video games and teenagers—Juvenile literature. | Internet—Safety measures—Juvenile literature. | Classification: LCC GV1469.3 .S778 2026 (print) | LCC GV1469.3 (ebook) | DDC 794.8/3—dc23/eng/20241121 | LC record available at https://lccn.loc.gov/2024054682